WHY THE WORLD LOOKS SO YOUNG

BRIAN THOMAS
WITH
JOHN MORRIS, JAKE HEBERT, & TIMOTHY CLAREY

Dallas, Texas
ICR.org

WHY THE WORLD LOOKS SO YOUNG
Scientific Evidences for the Genesis Creation and Flood

by Brian Thomas, Ph.D., with John Morris, Ph.D., Jake Hebert, Ph.D., & Timothy Clarey, Ph.D.

First printing: October 2020

Copyright © 2020 by the Institute for Creation Research. All rights reserved. No portion of this book may be used in any form without written permission of the publisher, with the exception of brief excerpts in articles and reviews. For more information, write to Institute for Creation Research, P. O. Box 59029, Dallas, TX 75229.

All Scripture quotations are from the New King James Version.

ISBN: 978-1-946246-54-7

Please visit our website for other books and resources: ICR.org

Printed in the United States of America.

TABLE OF CONTENTS

PAGE

Introduction 5

1. Why the Human Race Looks Young 7
   Population Growth: Earth Hit the Seven-Billion Mark
      Too Late 7
   Mitochondrial DNA: DNA Trends Confirm Noah's
      Family 9
   So Few Mutations: Human Mutation Clock Confirms
      Creation 11

2. Why the Fossils Look Young 15
   Soft Tissue: Mosasaur Retina Remnants, Blood Residue,
      and Skin Structures 15
   Soft Tissue: Duck-Billed Dinosaur Blood Vessels 17
   Bone: Dinosaur Bone Tissue Study Refutes Critics 18
   Radiocarbon: Carbon-14 Found in Dinosaur Fossils 19

3. Why the Earth Looks Young 23
   Continents: The Continents Should Have Eroded
      Long Ago 23
   Rock Layers: Flat Gaps Between Strata 24
   Rock Layers: Mount St. Helens 26
   Mantle Tomography: Cold Slabs Indicate Recent
      Global Flood 27

4. Why Outer Space Looks Young 31
   Big Bang Problems: Inflation Hypothesis Doesn't
      Measure Up 31

Galaxies: "Early" Spiral Galaxy Surprise.............32
Galaxies: Spiral Galaxy Model Leaves Questions
    Unanswered 34
Blue Stars: Study Can't Explain Blue Stragglers' Youth...36
Blue Stars: Young Blue Stars Found in Milky Way......38
Comets: Hartley 2............................39
Comets: NEOWISE 41

Conclusion 45

Introduction

What difference would it make to discover that real, measurable processes show a young world? This kind of science would support the Bible's teaching that God created the world in six days only 6,000 or so years ago. It would mean the Bible's history is spot-on. And for those who doubt Scripture based on secularists' non-biblical claims about the past, this evidence would elevate everything else the Bible says.

The Lord Jesus last walked on this earth about 2,000 years ago, Abraham lived about 2,000 years before Him, and the Bible's chronology from creation to Abraham took just over 2,000 years. It turns out the world—near and far, high and low—is packed with solid scientific evidence that aligns with this recent creation perspective. Why don't secular sources that shape our culture feature this evidence more openly? Where is its mention in nature shows, Hollywood movies, magazines, or college classrooms? Where can a person go to find and consider such science? This booklet meets that need.

Does God's world match God's Word when it comes to the history of our beginnings? This short collection of young-looking features ranging from our own bodies to outer space gives good reasons to favor the Bible's timeline. The Lord Jesus told Nicodemus, "If I have told you earthly things and you do not believe, how will you believe if I tell you heavenly things?" (John 3:12). Stated positively, if the Bible is right about where we came from and when, then we better pay close attention to what it says about where we're going!

1
Why the Human Race Looks Young

Population Growth: Earth Hit the Seven-Billion Mark Too Late

The world's population reached seven billion on October 31, 2011, according to the United Nations. Media outlets heralded the issue of overcrowding on the planet. How long did it take for this many humans to be born?

The evolutionary version of human population growth presents a fantasy scenario to answer that question. In this imaginary long-ages history, the population did not grow at all for millions of years before suddenly taking off only a few thousand years ago. In the July 29, 2011, issue of *Science*, demographic anthropology expert Jean-Pierre Bocquet-Appel wrote:

> After the members of the genus *Homo* had been living as foragers for at least 2.4 million years, agriculture began to emerge in seven or eight regions across the world, almost simultaneously at the beginning of the Holocene.[1]

Supposedly, the advent of agriculture enabled population growth at that time. But according to the Bible and historical records, there was never a time when humans weren't engaged in agriculture.

The problem is that in this projected timeline, people ("genus *Homo*") must have had virtually no population growth "for at least 2.4 million years." Bocquet-Appel wrote, "The world's population on the eve of the emergence of agriculture is estimated to have been 6 million individuals."[1] Thus, the first human couple that supposedly evolved from ape-like ancestors would have had only six million descendants after 2.4 million years. This requires a population growth rate of about 0.00029%—essentially zero. Virtually no growth for *2.4 million years*? If that really happened, where are all the remains of millions of years of pre-human bones? Instead, according to paleoanthropologist Lee Berger, we only have a few thousand small fragments—80% of which are isolated teeth, and "just a few dozen" skulls.[2]

In contrast, the average historically observed growth rate has been at least 0.4%, at times spiking to above 2%. Even a "pre-industrial farming population" growth rate of 0.1% per year—Bocquet-Appel's number—would have yielded today's seven billion people in only 7,062 years.[1] As ICR founder Dr. Henry Morris asked:

> How could it be that the planet only now is experiencing a population crisis—why not several hundred thousand years ago, soon after man first appeared on earth?[3]

To try to explain this slow growth, Bocquet-Appel stated, "An increase in the birth rate was closely followed in time by an increase in mortality." And the cause of all this death was infectious diseases such as "Rotavirus and Coronavirus."[1]

But this only invokes more unlikely events. How could such diseases maintain a near zero balance of birth and death rates for so long without randomly killing the whole population at some point? And why would these diseases suddenly lose their population-reducing effect after so many supposed eons? Why should they have been so much more virulent in the past than they are today? The survival rate for coronavirus COVID-19, for example, hovered around 97% of affected individuals across multiple countries during the 2020 pandemic—hardly a population reducer. Plainly, the infectious disease idea, along with unrealistically slow growth rates, are ad hoc add-ons that prop up long-age thinking.

The current world population aligns with biblical history with no added stories. Using census records from the last 400 years and a bit of algebra, and assuming a natural logarithmic growth, eight Flood survivors 4,500 years ago produce seven billion people in 2011 almost exactly.[4]

This powerful evidence points to accurate biblical history.

Mitochondrial DNA: DNA Trends Confirm Noah's Family

When research biologist Dr. Nathaniel Jeanson plotted hundreds of human mitochondrial DNA (mtDNA) sequences onto a tree diagram, the project revealed an obvious pattern: The mtDNA stemmed from three central "trunks" or nodes instead of just one. Three trends from Jeanson's data validate the biblical beginnings of mankind.

In order to understand just how, one must first recall that mothers pass mtDNA to every new generation. It comes from the mother's egg cell and contains 16,569 chemical base pairs—either adenine-thymine or guanine-cytosine—organized to encode vital information, like words in an instruction manual. Sometimes a DNA copying error known as a *mutation* leaves a different base in place of the original. Several empirical studies reveal that about one human mtDNA mutation occurs every six generations.[5] When a mother's egg cell mtDNA mutates in one place, the child conceived from that egg cell inherits that difference, along with later descendants if the child is female. This leaves a genetic trail that can lead back to mtDNA ancestry.

For his research, Jeanson downloaded mtDNA sequences taken from all major people groups. He then used standard software that arranges the most similar sequences closest together. The result is a tree-like diagram depicting lines of ancestry.

Jeanson's data show that the human mtDNA tree has three nodes.[5] Thus, everyone alive today carries one of three unique ancestral maternal sequences. This fits Genesis' claim that every human who exists today descended from one of the wives of Noah's sons.

Some perceive a fourth node in the lower right of the figure from Jeanson's paper. This area should not be considered a node because the distance between it and the nearest central node is almost twice as far as that between the three central nodes. This distance implies perhaps 20 generations, not the eight or so pre-Flood female generations inferred from the Genesis 5 genealogy. Several other regions are spaced just as far as the lower right region, yet they are not mistaken for nodes because

Why the World Looks So Young

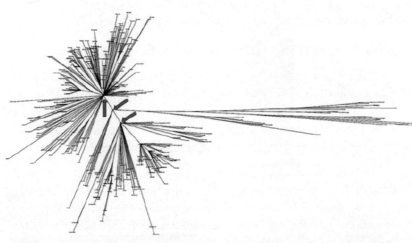

The human mitochondrial DNA tree shows three central nodes, marked by arrows. These fit the number of expected mDNA sequence differences between the wives of Shem, Ham, and Japheth. The longest branches represent the highest number of mDNA differences between people groups, and these numbers match Bible-based predictions. Image adapted from Figure 1 of Jeanson's paper.[5] Used by permission of Answers in Genesis.

other branches crowd more tightly around them. Only the three central nodes have the short branch lengths expected from the pre-Flood number of generations.

In other words, the distance between the three central nodes exactly confirms Genesis 5. At today's mtDNA mutation rate, two to eight nucleotide differences would have accumulated in the nine generations between Adam and Noah. The distance between the three central nodes also shows eight DNA differences.[6]

How many mtDNA differences would mutations cause during the 4,500 or so years since Noah? We need a mutation rate per generation and the total number of generations in order to find out. Jeanson noted that several empirical studies reveal about one human mtDNA mutation occurs every six generations. Generation times differ but fall within a tight range. At most, a culture where the women typically give birth near age 15 could have produced 115 mtDNA differences since Noah.[7] Adding the eight estimated pre-Flood differences totals 123. In a spectacular confirmation of Genesis history, the most diverse human mtDNA on record actually shows 123 differences.[8]

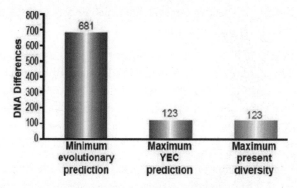

Predictions of mtDNA diversity independent of ethnicity. The minimum mtDNA diversity predicted by the evolutionary timescale is represented by the height of the left bar. The maximum mtDNA diversity predicted by the YEC timescale is represented by the height of the middle bar. Both predictions were compared to the maximum pairwise mtDNA difference observed and documented in the published literature (right bar). The YEC prediction clearly captured the maximum actual mtDNA diversity, but the minimum evolutionary prediction far exceeded this number. Image adapted from Figure 2 of Jeanson's paper.[5] Used by permission of Answers in Genesis.

In summary, if all peoples descended from three genetically unique mothers, then our mtDNA sequences should trace back to their three nodes. They do. Those nodes should have about eight differences between them. They do. Plus, a strict biblical timeline suggests 123 as the highest number of mtDNA differences that should be observed today. Check, check, and check. These three mtDNA trends trace all of humanity back to the three wives of Noah's sons—a striking intersection of biblical history and modern genetics.

So Few Mutations: Human Mutation Clock Confirms Creation

As mitochondrial DNA mutation rates have shown, human genetics strongly confirms the Bible's straightforward history. In studying mutation rates in the rest of human DNA—that found inside each cell's nucleus—geneticists have uncovered another clock-like countdown in human DNA. What does the fact that humanity's mutation clock is still ticking imply about the timing of human origins?

First, we contrast the evolutionary version of human history with the biblical version. Then we can evaluate which history best fits the clear implications from mutation studies.

The most widely accepted evolutionary conjectures assert that humanity evolved from an unidentified ape-like ancestor at least 2.4 million years ago. Humans may have experienced 120,000 generations in that supposed time.[9] In contrast, Scripture tells us there were about 100 generations from Adam to Christ.[10] Considering another 100 generations since Christ, a total of about 200 generations have passed since creation.

Now enter the human mutation clock. Geneticists use powerful new tools to directly compare DNA sequences between family members, and computers count every DNA difference, or mutation, that appears in each generation. Because neo-Darwinists consider mutations to be keys to the supposed evolution of humans from non-human ancestors, they are keen on tracking how fast mutations accumulate.

An extensive mutation rate study reported in *Nature* in 2012 resulted from tallying each specific mutation in 219 people, including 78 parent–offspring trios of Icelandic families.[11] It found an average of 63.2 new mutations per trio. This means each new generation suffers at least 60 new mutations. Mutations happen every generation when cellular processes fail to catch a handful of the 3.2 billion DNA "letters" found in each sperm or egg cell's DNA.

Past studies indicated that up to 10% of new mutations harm cells, most mutations cause no noticeable change, and beneficial mutations are virtually unknown.

Noted population geneticist Alexey Kondrashov reviewed the *Nature* study and agreed with its authors that these accumulating mutations very likely contribute to increasing incidences of diseases, including schizophrenia and autism. Additional studies confirm these claims.[12]

Plainly, human DNA sequence quality is relentlessly worsening. Kondrashov wrote:

> Because deleterious mutations are much more common than beneficial ones, evolution under this relaxed selection will inevitably lead to a decline in the mean fitness of the population.[13,14]

An inevitable decline in a population's "fitness" trends in the exact opposite direction to what evolution needs. Each new DNA typo represents a tick from a genetic clock counting down to zero. And everybody knows what happens to a clock that stops ticking.

Mutations in regular body cells damage cells within each one person's lifetime. Eventually, mutations render vital DNA sequences illegible to cellular machinery. Cell death results. When too few cells can cope, the whole body collapses under the weight of its degrading genome. And since mutations accumulate between generations, the same fate awaits the entire human race.

Nobody but the Creator can reverse this inevitable genetic decline. He is our only hope and offers exactly the kind of cure required—a whole new body that lasts forever. 1 Corinthians 15:21-22 say, "For since by man came death, by Man also came the resurrection of the dead. For as in Adam all die, even so in Christ all shall be made alive." Those who repent of their sins and trust in the Lord Jesus Christ will live with Him. Others will live totally apart from Him forever.

But for now, we live in a sin-cursed world. The mutation process sets a reasonable maximum theoretical limit to the total number of human generations. At the measured 60 new mutations per generation, evolution's 120,000 generations would produce 7,200,000 mutations among the three billion letters that comprise the human genome. This greatly exceeds the human genome mutation tolerance.[15] Without invoking a miraculous and extremely long suspension of mutational buildup, the human mutation rate alone precludes evolutionary history.

In contrast, the biblical estimate of 200 generations would have produced about 12,000 non-lethal mutations by now—enough to cause increasing diseases but not yet enough to ruin the human race. The mutational countdown is steady and relentless. The reason we have not yet reached the end must be because we began our journey recently—only thousands of years ago.

References
1. Bocquet-Appel, J.-P. 2011. When the World's Population Took Off: The Springboard of the Neolithic Demographic Transition. *Science*. 333 (6042): 560-561.
2. Google Earth and Human Evolution, Lee Berger, Talks at Google. Posted on youtube.com

November 22, 2012, accessed September 6, 2020.
3. Morris, H. 1984. *The Biblical Basis for Modern Science*. Grand Rapids, MI: Baker, 416.
4. The formula for logarithmic human world population growth is $P = P_0 e^{rt}$, where P = the current population, P_0 = the initial population, e = the base of natural logarithms (2.718), r = the average annual population growth rate (0.456% or 0.00456 in the calculator), and t = the time interval from P_0 to P.
5. Jeanson, N. T. 2016. On the Origin of Human Mitochondrial DNA Differences, New Generation Time Data Both Suggest a Unified Young-Earth Creation Model and Challenge the Evolutionary Out-of-Africa Model. *Answers Research Journal*. 9: 123-130.
6. Supplemental Table 4 ("Predictions of mtDNA differences under the YEC and evolutionary models") of the Jeanson paper shows an average of seven differences using a generation time of 35 years.
7. See Supplement Table 4, referenced above, for the derivation of this number.
8. Kim, H. L. and S. C. Schuster. 2013. Poor Man's 1000 Genome Project: Recent Human 5. Population Expansion Confounds the Detection of Disease Alleles in 7,098 Complete Mitochondrial Genomes. *Frontiers in Genetics*. 4: 1-13.
9. Assuming 20 years per generation.
10. Average the numbers of generations listed in Mary's genealogy from Luke 3 plus those given in Genesis 5 with the number of generations listed in Joseph's genealogy from Matthew 1.
11. Kong, A. et al. 2012. Rate of *de novo* mutations and the importance of father's age to disease risk. *Nature*. 488 (7412): 471-475.
12. Tennessen, J. A. et al. 2012 Evolution and Functional Impact of Rare Coding Variation from Deep Sequencing of Human Exomes. *Science*. 337 (6090): 64-69.
13. Kondrashov, A. 2012. The rate of human mutation. *Nature*. 488 (7412): 467-468.
14. Even tight selection, wherein the four least fit out of six hypothetical children die and only the two most fit survive, could never reverse mutational buildup since mutations are overwhelmingly non-beneficial and for other reasons. See Sanford, J. 2008 *Genetic Entropy and the Mystery of the Human Genome*, 3rd ed. Waterloo, NY: FMS Publications.
15. One study estimated human genome collapse after 32,800 years in a population of 1,011 and a 1.5% fitness decline per generation. See Williams, A. 2008. Mutations: evolution's engine becomes evolution's end! *Journal of Creation*. 22 (2): 60-66.

2
Why the Fossils Look Young

Soft Tissue: Mosasaur Retina Remnants, Blood Residue, and Skin Structures

Most mosasaur fossils are just a tooth or perhaps a broken rib or vertebra. Occasionally several bones are found still together. Conventional wisdom holds that creatures from Cretaceous rock layers died 70 or so million years ago. If that's true, why do some still have soft tissue?

The Los Angeles County Museum of Natural History has held the best-preserved mosasaur remains, found in Kansas in 1967, for decades. So much of the original mosasaur body remained intact that a study found details about its skin, eyes, and possibly its internal organs.

Paleontologist Luis Chiappe led a team of scientists publishing online in *PLoS ONE*. The researchers analyzed purple residue in the eyeball area of the mosasaur skull and concluded that it "may represent remnants of the retina."[1]

But soft tissues like this retinal tissue residue decompose even under ideal conditions. If this mosasaur was deposited "80 million years" ago,[1] why did its soft tissue remnants fail to turn to dust eons ago? The obvious implication of a more recent deposition went unreported in the report. If Noah's Flood buried this mosasaur only thousands of years ago, that could explain the tissue remnants.

Also found among the "exceptionally preserved soft tissue" were interesting dark red patches in the chest cavity. To find out what chemical stained the rocks red, the investigators submitted samples for chemical analysis—and the result was spectacular. They identified "hemoglobin decomposition products."[1] Hemoglobin is a major chemical constituent of blood, and biochemists know that it breaks down long before even one million years would elapse.

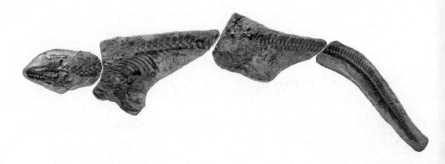

Image credit: Copyright © 2010 Lindgren et al, PLoS ONE 5 (8): e11998.

The researchers corroborated that the red color came from broken-down blood when they examined the positioning of the heart and liver within living ocean creatures. In dolphins and whales, these organs—as well as the lungs—are situated near the head to give their bodies a streamlined shape. One of the dark red patches in the mosasaur was right where a dolphin's heart would be located. It appears that these two blood-rich organs did not completely decay.

Despite the amazing find of the mosasaur's purple retinal and still-red blood tissues, "the most remarkable features of [this fossil] are the preservation of skin structures from all parts of the body."[1] The researchers were able to describe in detail the scale sizes and shapes almost from head to tail. Many of the small scales retained their three-dimensional shapes.

The very young-looking "wide range of soft tissue structures"[1] including skin, eyeball tissue, and blood-stained organ patches defy this fossil's 80-million-year age assignment. Considering that the Niobrara Chalk Formation that contained this mosasaur extends from the Gulf of Mexico to Manitoba, a recent worldwide flood should top the list of ways to explain Creataceous blood.

Soft Tissue: Duck-Billed Dinosaur Blood Vessels

Scientists keep finding short-lived biochemicals and even soft tissues in fossils. Over the years, they have found unmistakable evidence of specific proteins like collagen and hemoglobin, and even what look like red blood cells and bone cells, in dinosaurs and other fossils.[2] Most soft tissue structures occur as mineralized remains that preserve merely an impression or outline, but a few preserve decayed remnants of the original cellular structures. These original structures should be long gone after about one million years. A report of intact blood vessels in a duck-billed dinosaur bone pinpoints ways that such discoveries challenge old ideas about fossils.

A team of biomedical and earth scientists first chemically removed everything but the blood vessels from deep within the dinosaur bone.[3] They found 10 proteins, including tubulin, actin, myosin, tropomyosin, and histone H2A. A chemical analyzer read sequences of amino acids in each protein—like reading each word in an essay. They found enough similarities between the dinosaur proteins and those of modern reptiles and birds to conclude they were from a real animal, but enough differences to suggest that it was an extinct animal, like a dinosaur.

These dinosaur protein sequences help answer two key questions. First, did fungus or bacteria produce structures that masquerade as blood vessels? Well, bacteria don't even make these proteins. They're ruled out. And no known fungus makes hollow, branching tubes, so that rules them out, too. These results point straight to real dinosaur blood vessels.

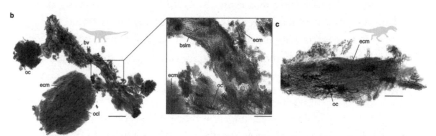

Partly polymerized soft tissues from demineralized Diplodocus *and* Allosaurus *bone. Images show remnants of osteocycles (oc), osteocycle lacunae (ocl), a blood vessel (bv), extracellular matrix (ecm), and basal lamina (bslm). Image credit: Wiemann, J. et al. 2018. Fossilization transforms vertebrate hard tissue proteins into N-heterocyclic polymers.* Nature Communications. *9: 4741.*

This leads to a second question: Can real blood vessels last 80 million years? Definitely not, and here's why. Scientists have not measured decay rates for these 10 specific dinosaur blood vessel proteins, but they have measured collagen and DNA decay. Repeated lab experiments show collagen should last no more than about a million years if kept cold.[4] Do these other proteins show scientific evidence that they could last many times longer than collagen given ideal conditions? Chemistry shows just the opposite.

For example, consider one of the duck-billed dinosaur proteins called *beta tubulin*. The team found it contained chemically stable amino acids like glycine and valine. These could last a very long time, but the protein's aspartic acid and methionine remain ready to react. Organic chemists have watched batches of aspartic acid react quickly and easily with available oxygen, and methionine readily reacts with oxygen to form methylsulfoxide. Why would these remain ready to react after millions of years?

These standard chemical breakdowns eventually turn blood vessels into gooey puddles. Scientists found methionine and un-reacted aspartic acid inside dinosaur blood vessels—not gooey puddles. Thus, even though we don't have specific decay rates for proteins like tubulin or actin, the amino acids that make them up—also found in collagen—should decay at least as fast as collagen.[5] Because no empirical study has demonstrated that a protein could last much beyond a million years, these researchers supported their belief in deep time with faulty logic instead of science.[6]

Blood vessels in dinosaur bones look young. Why? Because they are made of proteins with short-lived amino acids that look like they have not been around long enough to completely react with oxygen. This good science again confirms Scripture's account of a recent creation.

Bone: Dinosaur Bone Tissue Study Refutes Critics

Original dinosaur tissues in fossil bones are probably the most controversial finds in all of paleontology. Secular scientists have difficulty interpreting them. They debate whether the tissues are real based on laboratory-measured tissue decay rates, or whether the tissue decay rates are real based on plainly observed tissues.

A thorough report on this subject characterized original dinosaur biochemicals found inside fossil bones, and it adds further proof that these chemicals came from the long-buried animals themselves.

Many of those who are familiar with the rapid pace of tissue decay, which occurs as the tissues oxidize into tiny chemicals, have insisted that what looks like dinosaur blood vessels and cells is actually bacterial biofilm. Certain bacteria can produce slimy film structures as protective coatings. The study authors wrote in the journal *Bone*, "It has been proposed...that the [dinosaur bone] 'vessels' and 'cells' arise as a result of biofilm infiltration; but no data exist to support this hypothesis."[7]

The lead author of the study, North Carolina State University's Mary Schweitzer, used an array of different techniques to analyze the apparent bone cells inside the dinosaur bone. One method used antibodies, which are chemicals that bind to specific targets. She and her co-authors found that antibodies known to bind chemicals that vertebrates, not bacteria, produce clearly indicated that original vertebrate proteins were in the dinosaur bones.[8]

Thus, not only are there no data to support the biofilm idea, but there are plenty of data to outright disprove it.[9] Schweitzer and her colleagues wrote, "Here, we present morphological, microscopic, and chemical evidence that these are indeed altered remnants of original cells."[7]

So, the problem with finding soft tissues in dinosaur and other fossils only remains a controversy for secular scientists who will not yield, no matter what the evidence says, on their insistence that these artifacts are millions of years old. After all, without that unscientific assumption, all the actual science makes sense.

Radiocarbon: Carbon-14 Found in Dinosaur Fossils

Radioisotope science directly challenges the millions-of-years dogma scattered throughout the blockbuster movie *Jurassic World* and its sequels. A 2015 *Creation Research Society Quarterly* (CRSQ) article presented never-before-seen carbon dates for 14 different fossils, including dinosaurs. Because radiocarbon decays relatively quickly, fossils that are even 100,000 years old should have no measurable radiocarbon left in them.[10] But they do.

The CRSQ study authors tested seven dinosaur bones, including a *Triceratops* from Montana and hadrosaurids. They also sent fossils of a cartilaginous paddlefish, a bony fish, fresh-looking wood, and lizard bones from Permian layers in Canada and Oklahoma to radiocarbon laboratories. Five different commercial and academic laboratories detected carbon-14 in all the samples, whether from Cenozoic, Mesozoic, or Paleozoic source rocks. How did that radiocarbon get there?

The team also compared the results to several dozen published carbon-14 results for fossils, wood, and coal from all over the world and throughout the geologic column. Comparable amounts of radiocarbon showed up in almost 50 total samples so far described.[11]

Since radiocarbon can only last thousands of years, defenders of millions of years will have to assert that the radiocarbon came from contamination. These scenarios invoke recent or modern carbon that somehow crept into all these samples. This has been argued before, but the radiocarbon dating process itself is loaded with procedures that rigorously remove contaminants. No known source of radiation has enough energy to manufacture radiocarbon underground. Today, intense cosmic radiation does the job in the upper atmosphere.

What are the odds that imaginative contamination scenes could have occurred for one fossil, let alone for 50? Biased, glib claims of contamination weaken with every new documented carbon date from really old material.

Plenty of skeptics would discount all these data simply because they come from scientists who believe that Noah's recent Flood made most of Earth's fossils. Would it make a difference if secular scientists kept finding young-looking radiocarbon in Earth materials too?

In a study intended to refute original tissues, a team of evolutionary paleontologists carbon-dated a centrosaur.[12] The team reported radiocarbon in units called $F^{14}C$, which refers to the fraction of modern radiocarbon. Their result of 0.0149 translates to about 33,790 carbon years old.[13]

Another team found 24,600 carbon years' worth of radiocarbon in a mosasaur fossil.[14] Yet another—spectacularly—in a Cambrian sponge fossil supposedly 505 million years old. If it's that old, then why does it still have original biochemistry along with measurable radiocarbon inside it?[15] The Flood explains these results. It deposited all the rock layers from the lowest Cambrian to the highest Cretaceous and more—all during one world-destroying year only 4,500 or so years ago.

Jurassic World delivers good entertainment as long as viewers overlook its bad science. Real-world experiments counter repeated assertions of "millions of years" and see the much shorter timeline of Noah's Flood float to the top.

References
1. Lindgren J. et al. 2010. Convergent Evolution in Aquatic Tetrapods: Insights from an Exceptional Fossil Mosasaur. *PLoS ONE*. 5 (8): e11998.
2. Thomas, B. 2014. Original-Tissue Fossils: Creation's Silent Advocates. *Acts & Facts*. 43 (8): 5-9.
3. Cleland, T. P. et al. 2015. Mass Spectrometry and Antibody-Based Characterization of Blood Vessels from *Brachylophosaurus canadensis*. *Journal of Proteome Research*. 14 (12): 5252-5262.
4. Buckley, M. and M. J. Collins. 2011. Collagen survival and its use for species identification in Holocene-lower Pleistocene bone fragments from British archaeological and paleontological sites. *Antiqua*. 1 (1): 1-7.
5. A separate study showed that blood puree preserved vessels at room temperature for several years. Although this verified that iron from the blood inhibited microbes that degrade blood vessels, it inadvertently showed that iron failed to convert blood vessel proteins into a more time-resistant chemical. See Thomas, B. Can Iron Preserve Fossil Proteins for Eons? *Creation Science Update*. Posted on ICR.org June 23, 2015, accessed December 8, 2015.
6. Cleland et al wrote, "Application of these approaches to other fossil specimens, derived from both Mesozoic and Cenozoic deposits, will further identify the types of proteins likely to persist into deep time, or, more specifically, will identify particular functional groups or molecular characteristics that increase preservation potential." Well, of course these proteins last through "deep time" if one assumes they were buried 70 million years ago. But this circular argument ignores powerful evidence that blood vessel proteins could never last that long.
7. Schweitzer, M. H. et al. 2013. Molecular analyses of dinosaur osteocytes support the presence of endogenous molecules. *Bone*. 52 (1): 414-423.
8. Specifically, antibodies reacted with non-bacterial proteins actin, tubulin, PHEX, and histone H4.
9. See also Peake, T. Small Foot, Big Impression. *Phys.org*. Posted on Phys.org July 24, 2007, accessed October 26, 2012.
10. Thomas, B. and V. Nelson. 2015. Radiocarbon in Dinosaur and Other Fossils. *Creation Research Society Quarterly*. 51 (4): 299-311.
11. Every one of the 14 specimens had radiocarbon years BP (before present) of between 17,850 and 49,470, corresponding to evolutionary age assignments of 10 million and 280 million years respectively. Radiocarbon years do not correspond to calendar years because of past changes in the rate of carbon-14 formation and carbon content.
12. Saitta, E. T. et al. 2019. Cretaceous dinosaur bone contains recent organic material and provides an environment conducive to microbial communities. *eLife*. 8: e46205.
13. It's no problem to adjust these carbon years to biblical calendar years by, for example, factoring in a much larger pre-Flood biomass (which would dilute $F^{14}C$) and a much stronger

pre-Flood magnetic field (which would reduce ^{14}C production), but it's a huge problem to reconcile these carbon years to tens or hundreds of millions. See Cupps, V. R. 2017. Radiocarbon Dating Can't Prove an Old Earth. *Acts & Facts*. 46 (4): 9.
14. Lindgren, J. et al. 2011. Microspectroscopic Evidence of Cretaceous Bone Proteins. *PLoS ONE*. 6 (4): e19445.
15. Ehrlich, H. et al. 2013. Discovery of 505-million-year old chitin in the basal demosponge *Vauxia gracilenta*. *Scientific Reports*. 3: 3497.

3
Why the Earth Looks Young

Continents: The Continents Should Have Eroded Long Ago

According to standard evolutionary models, the earth is supposed to be 4.5 billion years old, and its continents supposedly formed 3.5 billion years ago. But if this is true, why haven't Earth's landforms been completely eroded and deposited into the seas?

Geologists have measured erosion rates in different ways. Once measured, we can extrapolate those rates into the past as a kind of test between evolution's required long ages and the Bible's record of recent beginnings. Results from one thorough study indicate that the earth's overall erosion rate, although slow, would have leveled the continents at least 70 times over if they are as old as the evolutionary claim maintains!

Geologists have been measuring quantities of ^{10}Be, an isotope of the element beryllium that becomes radioactive with exposure to the sun. The more ^{10}Be present at a given site, the longer it has been exposed to the sun without being carried off by erosion. This system was used by dozens of geologists to estimate erosion rates around the world.

The Geological Society of America study showed the collated ^{10}Be data from every continent.[1] The researchers painstakingly converted all the reports to directly comparable units and found that erosion occurs 18 times faster in drainage basins than it does in outcrops.

According to the study, the average erosion rate for outcrops was 40 feet every one million years. The average thickness of continental crust above sea level can be estimated at about 623 meters, or 2,044 feet.[2] To erode 2,000 feet of crust at 40 feet per one million years would require only 50 million years. So, if the earth is *billions* of years old, why is its surface not completely flat?

Continental basins, or low-lying areas that receive more rainfall from higher elevations, erode much faster. Applying this rate, continents would have eroded down to sea level in less than three million years. So, the data in this study indicate that the overall time needed for continents to erode lies between three and 50 million years, a range that includes the 2007 estimate by ICR scientist Dr. John Morris of 14 million years.[3]

In order to maintain a belief in long ages, some might suggest that landforms were repeatedly uplifted by tectonic forces, providing more land mass for weather to erode. However, Loma Linda University geologist Ariel Roth noted that this scenario would have obliterated the very rock layers that supposedly represent evolution's millions of years. He wrote:

> It has been suggested that mountains still exist because they are constantly being renewed by uplift from below. However, this process of uplift could not go through even one complete cycle of erosion and uplift without eradicating the layers of the geologic column found in them. Present erosion rates would tend to rapidly eradicate evidence of older sediments; yet these sediments are still very well-represented, both in mountains and elsewhere.[2]

These reported erosion rates confirm this longstanding argument. The fact that mountains and even continents still exist is testimony to the young age of the earth. It looks as though the continents cannot be billions of years old because they would all have eroded in a fraction of that time. And yet they still stand tall.

Rock Layers: Flat Gaps Between Strata

When geologists make field observations, they typically focus on the rock before them and its color, density, mineral makeup, fossil content, and other features. But the strata's context would tell them even more about rock layers' origins.

Numerous examinations of local outcrops can result in large-scale maps and cross-sections. Both small-scale and large-scale studies are necessary, but big-picture consideration of the strata and timing of deposition help explain both the rock and the conditions under which it was formed.

Geologists have found that the layers come in "packages" of strata called *megasequences* that are bounded on top and bottom by evidence of erosion. The depositional package of sediments overlies a recognizable unconformity or erosional plane and is truncated at the top by another unconformity. Geologists identify at least six megasequences that together comprise essentially all sedimentation that happened on Earth. The packages persist across the continents, often ignoring the standard geologic column and yet fitting in with the megasequences. Nearly all sediments were either water-deposited or water-eroded. Could this be the signature of the global Flood?

The accompanying chart illustrates the various layers (shaded), the erosional unconformities (wavy lines), and the strata that are assumed to be missing either through erosion or non-deposition (cross-hatched).[4] Such charts could be drawn anywhere, but the well-studied and well-represented layers in southern and eastern Utah illustrate the principle of flat layers. The chart shows many pancake-like sedimentary layers in sequence and the erosional gaps between them. The strata are plotted according to the dates assigned by standard thinking of their upper and lower surfaces. The layers actually lie directly on top of each other, but they are drawn separated in time. More of the total geologic column proposed by uniformitarianism is missing than is present.

For evolutionary geologists, the fact that layers are missing is evidence of erosion. But obvious evidence for erosion is missing as well. Evolutionists assign the time between two layers as tens of millions of years, but the contacts are typically flat and featureless. Millions of years of erosion would produce irregular terrain, but there is none—no streambeds, valleys, or canyons.

Road cuts often reveal flat and featureless contacts between strata. Some extend for many miles. The big-picture stratigraphic sections, however, reveal flat time gaps that span the continent, creating doubt about the passage of a long time period and implying dynamic floodwaters.

Why the World Looks So Young

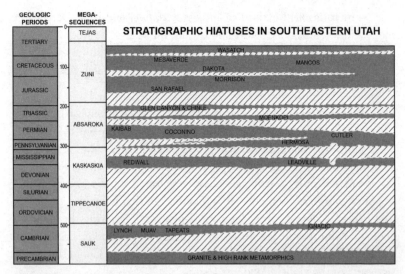

The sedimentary layers in southeastern Utah shown in proper time relationship (listed in millions of years per uniformitarian thinking). The shaded areas represent strata, while the cross-hatched areas are assumed time gaps. The standard geologic column is given on the left. The horizontal distance represents about 200 km, while the total thickness of the actual strata is about 3½ km. Actually, the strata are resting on top of each other with no gaps between. The time gaps are required by standard thinking regarding the geologic "ages."

These types of discussions were never held a generation ago, but expanding geologic knowledge has made regional maps and other data available. No longer should geologists restrict their focus to a single outcrop or hand specimen while ignoring larger implications. No longer should catastrophic thinking, including what the Genesis Flood would have done, be excluded.

Rock Layers: Mount St. Helens

Washington State's Mount St. Helens erupted catastrophically in 1980. Its first burst filled an entire valley with hundreds of feet of sediment. Another smaller eruption deposited more material on top of that, and then a third deposit occurred in 1982. Later, a small volcanic burp melted the glacier that had built up in the crater. Resulting snowmelt tore a gash through those fresh deposits, revealing sharp and flat contacts between each of them. A close-up look showed that fast-flowing currents can lay down multiple layers thinner than a finger width.

Mount St. Helens revealed that both thick and thin layering can happen fast. Millions of years aren't needed to form sedimentary rock or stratigraphic layering. Sedimentary layers hundreds of feet thick formed within hours during the eruption itself and then hardened into rock soon after the water drained from them. What was once muddy debris flying faster than cars on a freeway stiffened into rocks strong enough to form steep cliffsides in no more than two years. Could other layered sedimentary rocks in Earth's crust have formed rapidly?

Because of the Mount St. Helens eruption, scientists know that sedimentary rock layers and steep-walled canyons can form in only hours. Fast layering, fast hardening, and fast canyon carving from Mount St. Helens make the world's layers, rocks, and canyons look young. One doesn't need vast time to explain these Earth features—just vast water. Of course, Scripture told us all along about this water, saying, "For this they willfully forget: that by the word of God the heavens were of old, and the earth standing out of water and in the water, by which the world that then existed perished, being flooded with water" (2 Peter 3:5-6).

Mantle Tomography: Cold Slabs Indicate Recent Global Flood

Since the 1990s, cross-section images of mantle tomography have shown visible slabs of oceanic lithosphere (which includes oceanic crust) descending hundreds of miles beneath ocean trenches into subduction zones.[5] These descending plates have been imaged all the way down to the top of the earth's outer core[6] and are composed of cold, brittle, dense rock about 62 miles thick.

Researchers from the University of Colorado reported finding that some of the subducted slabs stagnate at depths of about 670 km (416 mi) to 1,000 km (620 mi) and appear to travel horizontally. Publishing in *Nature Geoscience*, Wei Mao and Shijie Zhong presented seismic tomography that shows descending plates beneath the Honshu and northern Mariana subduction zones in the western Pacific. The plates stop in the mantle transition zone and move horizontally for more than 1,500 km (930 mi) to the west beneath East Asia.[7]

Their research goal was to determine why some slabs extend to the core while others stop descending partway through the mantle and move horizontally. They concluded that a mineral phase change from increasing temperatures and pressure in the mantle transition zone is likely responsible, stating:

> We demonstrate that the observed stagnant slabs in the transition zone and other slab structures in the lower mantle can be explained by the presence of a thin, weak layer at the phase change boundary that was suggested by mineral physics and geoid modelling studies.[7]

The authors added that these stagnant plates seem to be a recent phenomenon occurring within the last 20 to 30 million years. But if these subducted slabs are really many millions of years old, why do they still show such prominent temperature anomalies? All of the images of the subducted slabs show consistently cooler rock surrounded by extremely hot mantle, even after traveling more than 1,500 km (930 mi) right through the mantle itself.[7] These rock slabs appear to be at least 1,000°C cooler than the surrounding mantle material at these depths, based on their density.[8]

And a bigger problem was never addressed in the paper. How do the cold slabs that extend down to the base of the mantle (at the core boundary) remain cold after 30 to 50 or more million years at the sluggish subduction rates secular scientists envision? These lithosphere slabs had to travel 2,900 km (1,800 mi) to reach the base of the mantle where the temperature is even hotter, about 3,500°C—over 6,300°F.

The cooler temperatures exhibited by the subducted slabs create a thermal dilemma for secular and old earth geologists, who must explain how these slabs remained cold for tens of millions of years. Instead, these cold lithospheric slabs indicate they were rapidly emplaced just thousands of years ago.

Cold subducted slabs are best explained by runaway subduction. The catastrophic plate tectonic model suggests that subduction during the Flood year occurred rapidly, with plates moving at rates of meters per second—several miles per hour.[6] These fast speeds could easily place long slabs of cold lithosphere deep within the hot mantle as we observe today. Since the slabs have only been within the mantle for a few thousand

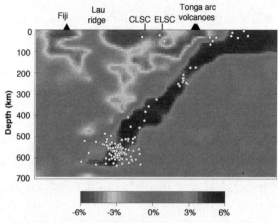

P-wave tomography under the Tonga Trench, Pacific Ocean. The darkest gray shows the colder ocean lithosphere descending down into the mantle to a depth of nearly 700 km (435 miles). The white dots represent earthquake foci. Image credit: Copyright © American Association for the Advancement of Science. Adapted for use in accordance with federal copyright (fair use doctrine) law. Usage by ICR does not imply endorsement of copyright holder.

years, they are still much cooler than the surrounding mantle. Today's measured slow plate speeds are what was "left over" after runaway subduction ended.

If these lithospheric slabs and those at the base of the mantle were really moving as slowly as secular scientists claim at just a few centimeters per year, then the slabs should have warmed up and assimilated long ago and not show such strong density contrasts—indicating a much-cooler temperature—with the surrounding hot mantle.

Dr. Jake Hebert aptly summarized the findings from mantle tomography:

> An imaging process called *seismic tomography* has revealed a ring of dense rock at the bottom of the mantle. Since its location corresponds approximately to the perimeter of the Pacific Ocean, it appears to represent subducted ocean crust. Located inside this ring of cold rock is a blob of less-dense rock that appears to have been squeezed upward toward the crust. If one assumes that the density of the cold ring is comparable to that of the surrounding material, which is

the most straightforward assumption, this ring is 3,000 to 4,000°C colder than the inner blob. This is completely unexpected in the conventional plate tectonic model since it can take about 100 million years for a slab to descend all the way to the base of the mantle. In that time, one would expect any such temperature differences to have evened out. However, in the catastrophic plate tectonics model, such a temperature difference is to be expected if the slab rapidly subducted into the mantle just a few thousand years ago.[8]

Mantle tomography showing blobs of cold subducted rock confirm the research findings of creation scientists, validating runaway subduction.[9]

Rapid plate motion only occurred during the Flood year about 4,500 years ago. Once the original colder oceanic lithosphere was completely consumed by subduction and a new hotter seafloor was produced, the runaway subduction process ceased. Today, we witness only residual plate motion. Seismic tomography images showing cold subducted plates deep in the mantle remind us that these events took place in the time frame of the Bible.

References
1. Portenga, E. W. and R. R. Bierman. 2011. Understanding Earth's eroding surface with ^{10}Be. *GSA Today*. 21 (8): 4-10.
2. Roth, A. A. 1986. Some Questions about Geochronology. *Origins*. 13 (2): 64-85.
3. Morris, J. D. 2007. *The Young Earth*. Green Forest, AR: Master Books, 93.
4. Diagram modified from Roth, A. A. 1988. Those Gaps in the Sedimentary Layers. *Origins*. 15 (2): 75-92.
5. Grand, S. P. et al. 1997. Global tomography: a snapshot of convection in the Earth. *GSA Today*. 7: 1-7.
6. Baumgardner, J. R. 2003. Catastrophic Plate Tectonics: The Physics Behind the Genesis Flood. In *Proceedings of the Fifth International Conference on Creationism*. R. L. Ivey, Jr., ed. Pittsburgh, PA: Creation Science Fellowship, 113-126.
7. Mao, W. and S. Zhong. 2018. Slab stagnation due to a reduced viscosity layer beneath the mantle transition zone. *Nature Geoscience*. 11: 876-881.
8. Hebert, J. 2017. The Flood, Catastrophic Plate Tectonics, and Earth History. *Acts & Facts*. 46 (8): 11-13.
9. Baumgardner, J. 1994. Runaway Subduction as the Driving Mechanism for the Genesis Flood. In *Proceedings of the Third International Conference on Creationism*. R. Walsh, ed. Pittsburgh, PA: Creation Science Fellowship Inc., 63-75.

4
Why Outer Space Looks Young

Big Bang Problems: Inflation Hypothesis Doesn't Measure Up

Since the Big Bang story of the origin of the universe has been refuted by a host of external observations and internal contradictions,[1] secular science has been forced to postulate additional, exceedingly improbable events to keep it afloat. One of these is *inflation*, which attempts to explain the apparent uniformity of the universe.[2] But observations by the Wilkinson Microwave Anisotropy Probe are forcing cosmologists to revamp inflation, at the cost of inventing yet another miraculous event to prop it up.[3]

Inflation holds that 10^{-36} seconds after the Big Bang was ignited, the expansion rate of the universe increased by a thousand-billion-billion-billion-fold. It was supposedly an explosive event that followed immediately after the larger explosive event. This hypothesis, however, fails to address critical deficiencies in the overall Big Bang model.

Supposedly, inflation miraculously provided the force necessary to keep the newly ignited material of the universe from immediately collapsing upon itself.[4] Inflation also attempted to solve the horizon problem—how the universe could have achieved the incredibly uniform temperature that it has in only 13.8 billion years when at least twice this length of time would have been necessary.[5]

The universe had to expand at a certain rate to prevent self-collapse, but this rate is many times the speed of light. Contrarily, the nascent universe must have expanded slower than the speed of light in order for that light to have had enough time to bathe every "corner," for only in this way could the temperature be so consistent throughout today's universe. But this means the universe must have expanded faster than it physically could have, rendering inflation flat. Two additional problems with inflation include "(a) how to get it started, and (b) how to stop it."[6]

Measurements of the low-energy background radiation from space, the cosmic microwave background radiation (CMBR), add even more complications. Half of the universe has temperature and density variation patches, and the other half is quite smooth. According to the California Institute of Technology scientists who published their Wilkinson Microwave Anisotropy Probe observations:

> Such an asymmetry cannot be generated during single-field slow-roll inflation without violating constraints to the homogeneity of the Universe.[3]

In other words, the old inflation hypothesis cannot account for the lopsided universe, if indeed these CMBR measurements are accurate.

Thus, Caltech astrophysicist Sean Carroll offered "a new version of inflation theory" in which one "field" was "responsible for ballooning the size of the universe, while a second field...introduces the density variations."[7] But this additional field comes with a cost. It "requires the introduction of a large-amplitude superhorizon perturbation to the curvaton field."[3] This perturbation structure, Carroll proposed, would have existed prior to inflation, but it only adds another ad hoc assumption to Big Bang's big list. Where would the specifications of this proposed perturbation have originated?

Carroll's effort to reconcile this monumental asymmetry of the universe, albeit into a woefully inadequate cosmology, is at least some kind of attempt to reckon with the data. Cosmologies other than, and better than, Big Bang are being investigated, and several of them predict the young world described in the Bible. The information that can actually be measured from the universe soundly contradicts the Big Bang's series of miraculous natural events,[8] but it fits well with a biblically consistent view of the structure and origin of the universe.

Galaxies: "Early" Spiral Galaxy Surprise

The naked eye allows us to see just a little of God's heavenly handiwork, but even this little bit clearly declares God's glory.[9] As more powerful telescopes peer deeper into space, more and more reasons to question the secular origins story accumulate.

Astronomers were surprised by very distant spiral galaxies. As described in a popular science news article, "New results from an ambitious sky survey program, called ALPINE, reveal that rotating disk-shaped galaxies may have existed in large numbers earlier in the universe than previously thought."[10] These rotating galaxies were detected using data from multiple observatories, including Chile's Atacama Large Millimeter/submillimeter Array, or ALMA. By observing Doppler shifts[11] of light emitted from positively charged carbon ions within the galaxies, astronomers were able to infer that the galaxies were rotating. By secular reckoning, these galaxies would have formed between one and 1.5 billion years after the supposed Big Bang, or 12.2 to 12.7 billion years ago.[10]

Big Bang astronomers assume that light from the most distant galaxies takes billions of years to reach Earth. Indeed, this assumption is often used to challenge Bible believers. It is at the heart of the question "If the universe is just thousands of years old, how can we see starlight from galaxies that are billions of light-years away?"

However, Einstein's theories of relativity have shown that supposedly intuitive ideas about light, space, and time are incorrect, and many creationists think that relativity theory may be the key to answering this common objection to biblical creation.

Moreover, this assumption is a double-edged sword for secular astronomers because distant light presents problems for the Big Bang, too.

Because Big Bang astronomers assume light from the most distant galaxies takes billions of years to reach Earth, they think we are seeing these very distant galaxies not as they are now but as they were in the very distant past. Since galaxies supposedly take vast amounts of time to form and evolve, the most distant galaxies should appear to us to be immature and poorly developed. However, many galaxies that are extremely distant appear fully formed or mature![12-15]

Better observations will be necessary to confirm if these disk-shaped objects are true spiral galaxies. If so, this would turn out to be just one more example of this phenomenon.[16]

ICR has numerous book and video resources available on astronomy-related topics. These include our four-episode DVD series *The Universe: A Journey Through God's Grand Design*, which is also available via digital download.[17] We also carry all three volumes of the highly popular DVD series *What You Aren't Being Told About Astronomy*,[18] as well as astronomy- and space-themed books including *Guide to the Universe*,[19] *The Work of His Hands* by astronaut Jeff Williams,[20] and a number of books for younger readers.[21-23]

Galaxies: Spiral Galaxy Model Leaves Questions Unanswered

According to evolutionary astronomers, the arching "arms" that protrude from spiraling galaxies are supposedly 10 billion years old.[24] But because the inside stars rotate faster than the outside stars, the arms they form would have "wound up" to form a disk after only about 100 million years. And what possible forces could have spun them out of a mass of hydrogen atoms or a clump of stars in the first place?

Postgraduate student Robert Grand of the Mullard Space Science Laboratory, University College London, presented a spiral galaxy model at the Royal Astronomical Society's National Astronomy Meeting in Wales. The model attempted to digitally simulate the origin of the arms in spiral galaxies.

Grand said in an RAS press release, "We have found it impossible to reproduce the traditional theory, but stars move with the spiral pattern in our simulations at the same speed."[25] The traditional theory held that stars are channeled into arms by "density waves." However, this explanation merely pushes the problem back one step to another puzzle—what caused the density waves? And clearly, Grand was unable to make the physics of the traditional theory of density waves work in his simulations.

His replacement model appears to be based on spiral density waves, but his version, used to model a Milky Way-size galaxy, showed how stars could rotate within each galactic arm. After 100 million years, "the arm breaks up due to the shear forces."[25] This is consistent with what physicist Russell Humphreys wrote in 2005 when he briefly described the winding galaxy problem:

The stars of our own galaxy, the Milky Way, rotate about the galactic center with different speeds, the inner ones rotating faster than the outer ones. The observed rotation speeds are so fast that if our galaxy were more than a few hundred million years old, it would be a featureless disc of stars instead of its present spiral shape.[24]

Thus, neither the traditional density waves nor this spiral density wave model show that the spiral arms could last anywhere near the 10 billion years that evolutionists have assigned to many distant galaxies.

But Grand did suggest an escape from this problem. His presentation abstract stated, "We find that the spiral arms are recurrent material features, and the pattern speed generally decreases with radius, in such a way that the pattern speed almost equals the rotation speed of stars at all radii."[26]

Though the spiral arms may be "recurrent" in his simulations, real spiral galaxies within the universe do not show stages of arms forming, collapsing, and reforming. Actual galaxies either have strongly defined spiral arms, like the ones in Galaxy M81, or arms that have partly or totally blurred into one another. There are no examples in the skies of new arms forming, and this model did not produce the clear spiral arms that telescopes have revealed.

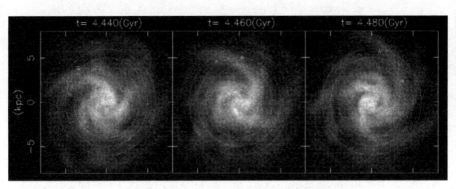

Image credit: Robert Grand. Copyright © 2011 Royal Astronomical Society.

Humphreys noted this mismatch between the way real galaxies look and the virtual galaxies in Grand's model, shown above.

In a relevant email, he wrote:

> I notice that the spirals of their simulations are very poorly defined, with not much fewer stars per unit area between the arms compared to the number per unit area in the arms, and very blurry boundaries. But many real galaxies have very clearly defined spiral arms.[27]

Humphreys also noted that without access to the details of Grand's model, it is impossible to discern if the claim of rapid spiral arm formation is the product of wishful thinking, unrealistic tweaking of physics parameters, or real physics. But since Grand's model produced such a weak facsimile, there is little reason to trust it.

Thus, a realistic way for natural processes alone to produce spiral arms has yet to be discovered. And spiral galaxies still don't look anywhere near as old as the "nature is all there is" advocates claim. Spiral galaxies still look like relatively youthful, intentional creations.

Blue Stars: Study Can't Explain Blue Stragglers' Youth

Blue stragglers, according to NASA, "are older stars that acquire a new lease on life when they collide and merge with other stars."[28] But one study called into question whether stellar collisions can account for these remarkable stars. And blue stars burn their fuel so quickly that they actually look young.

Since their discovery, evolutionary astronomers have sought a way to explain how these stars can even exist. They burn fuel so fast that they should have burned out billions of years ago. American astronomers Aaron Geller and Robert Mathieu published in *Nature* a description of their model for how older stars could have acquired a "new lease on life" by siphoning matter from nearby gas giant stars through "mass transfer."[29]

They investigated stars within a cluster called NGC 188, found in the constellation Cepheus. It contains 21 blue stragglers, 16 of which are binary stars that closely interact with nearby stars. The researchers suspected that the blue stragglers' partners were white dwarfs, which would be small, leftover remnants of larger red stars that the blue ones had drained of fuel. Such dwarfs are too faint for direct observation, but they have sufficient mass to cause their partner stars to wobble.

Of the 16 binaries, 12 had rotational periods right at 1,000 days and were thus called "long-period" blue stragglers. The study authors ran a statistical analysis that showed "the theoretical and observed [mass] distributions are indistinguishable."[29] In other words, their theory that other stars "fed" these 12 blue stragglers matched well with what they observed.

But did this reconcile the relative youthfulness of these binary blue stars with their assumed billions of years of history? The answer is no. The authors wrote, "Blue straggler stars…should already have evolved into giant stars and stellar remnants,"[29] and their new observations do not solve this deep-time problem.

Blue stragglers should burn through all their fuel in "a few million years at best."[30] But these NGC 188 stars are supposed to be seven billion years old. So, to make them fit that age, these authors maintained that they were not initially blue stars but instead burned fuel at a normal rate for billions of years. Then suddenly, within the last one million years, all 12 of them began siphoning extra fuel from their binary partners so that they only *look* young right now.

However, nothing explains the many blue stragglers that are not binary stars and yet exist near and far throughout the universe. Could they have received recent "youthfulness" through collisions with other stars?

In an article summarizing the Geller and Mathieu paper, University of Cambridge astronomer Christopher Tout wrote, "Thus, in this cluster [NGC 188], a collisional origin for blue stragglers is much rarer than expected, and the authors' study casts doubt on whether it occurs at all."[31] The modeled scenario in which blue stars form by collision did not match observations.

So, isolated blue stars could not have received their young looks from a binary system since by definition they have no binary from which to siphon fuel. They probably didn't receive their youthful appearance from collisions, either, according to these results. And though the binary blue stragglers may have siphoned fuel from nearby partners, the idea that 12 of 16 only did so recently—after an imagined seven-billion-year wait—defies reasonable odds.

Thus, the best explanation is still the most straightforward one—blue straggler stars look young because they *are* young.

Blue Stars: Young Blue Stars Found in Milky Way

The Hubble Space Telescope, which had been programmed to search for planets, found 42 "oddball" blue stars in the Milky Way galaxy. These stars burn so brightly that they consume their fuel much faster than other stars. Though they are found in more abundance in more distant galaxies, the discovery of nearby blue stars presents a particular problem for standard long-age cosmologies.

Blue stars should not exist in a universe that is 13.7 billion years old because they should have burned out billions of years ago. Retired University of South Carolina astronomer Danny Faulkner noted, "In fact, the hottest blue stars could last only a few million years at best. Both creationists and evolutionists acknowledge this fact."[32] Thus, evolutionists have proposed that these stars have been constantly generated during this long timespan. But that means blue stars should be forming even now. "Despite their diligent search, however, [astronomers] have never observed one of these blue stars forming—or any other star, for that matter," Faulkner wrote.[32]

The physical barriers to star formation by collapsing gases—the standard nature-only story of star formation—are prohibitive because the denser a cloud of gas becomes, the more vigorously its particles repel one another.[32] Thus, long-age astronomers have been forced to resort to a non-explanation that is indistinguishable from an appeal to magic—that blue stars somehow formed in the distant past through unknown, unobserved processes. A NASA news release announcing the Hubble discovery admitted, "It is not clear how blue stragglers form."[33]

They are considered "stragglers" because they appear to lag behind the aging process of their apparently older and redder-colored stellar peers. Also, according to the news release, some of the 42 objects may not be blue stars but "a mix of foreground objects," 18 to 37 of which are likely genuine blue stragglers.[33] The telescope was trained onto the galactic core, which is very dense with stars and gases.

Over the years, astronomers have observed more blue stars farther away and therefore further back in time. Thus, faith-based assertions that these short-lived stars resulted from natural, though unknown, formation mechanisms have been able to hide behind a shroud of distance. If an event occurred billions of years ago, then perhaps claims of "unknown processes" don't sound so crazy.

But these newly discovered blue stars are only 26,000 light-years away.[34] If these stars were products of nature and not God, then a reasonable nature-only explanation for how they formed should be straightforward. But it is not. Young, hot blue stars are very close (as well as very far away), so the lack of any reasonable natural explanation for their origin can no longer be hidden behind great distances. On the other hand, if the universe and its stars are only thousands of years old, then blue stars need no special explanation.

The scientific evidence shows, and the Bible clearly states, that blue stars were put in place on purpose recently.[35]

Comets: Hartley 2

Comet Hartley 2 is an odd, dumbbell-shaped object that rotates as it tumbles along its orbit. One end spews carbon dioxide gas so violently that it regularly throws off chunks of ice as it travels around the sun every six and a half years or so. Astronomers are scratching their heads over how such a small object could still have so much energy and material after billions of years of existence.

NASA's EPOXI mission flew near enough in November 2010 to capture impressive images of Comet Hartley 2 ejecting its gases. Much of the data it collected has since been analyzed. The University of Maryland's Michael Ahearn told *Space.com*, "Among the comets visited by spacecraft, Hartley 2 is in a class by itself."[36]

First, the comet looks young. *Space.com* reported:

> For starters, its nucleus contains an abundance of carbon dioxide (CO2—or, in solid form, dry ice). This is a volatile material—it burns easily—and so scientists would expect much more of it to have burned away in the 4.5 billion years since the comet formed along with the rest of the solar system.[36]

In fact, as ICR pointed out when data from the flyby were fresh, one would expect not just "much more of it" to have burned off but all of it eons ago.[37]

A detailed analysis of Comet Hartley 2 was published in the June 17, 2011, issue of *Science*.[38] A'Hearn, the study's principal investigator, emailed *Space.com*:

> We are still trying to sort out the implications for formation. The biggest remaining question, or at least the one that interests me most, is why there is so much CO2 in this comet and why it seems to differ between the two ends.[36]

So, nobody knows how this one-mile-long flying space rock could possibly still have so much carbon dioxide after billions of years. But this is a familiar story in the world of astronomy. Nobody knows how the material that supplies Saturn moon Enceladus' jet-like fountain could possibly still exist, either.[39]

The comet also shows an "excited state" of rotation, including a "nodding motion" that was unexpected. The authors of the *Science* re-

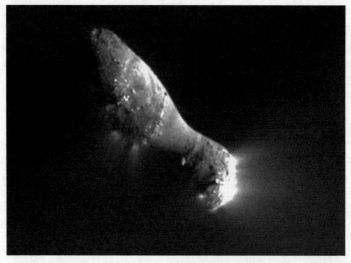

Secular scientists were shocked to find that Comet Hartley 2 is still outgassing carbon dioxide, even though it is supposedly billions of years old. Image credit: NASA/JPL-Caltech/UMD

port stated, "These changes are presumed to be due to torques produced by the outgassing."[38] But after 4.5 billion years, how could the comet's outgassing still be able to generate enough force to push around this 300-megaton comet?[40] The whole thing should have fizzled into space dust in only a fraction of its supposed billions-of-years lifespan.

Comet Hartley 2 looks very young. The most "scientific" explanation should be the one that fits the most data with the fewest assumptions. If it *looks* young, maybe it *is* young—even if its maximum age has to be measured in thousands, not billions, of years.[41]

Comets: NEOWISE

Stargazers were disappointed in early 2020 by comets ATLAS and SWAN, which disintegrated before they could put on good celestial shows. But another comet appeared in the sky and delighted stargazers.[42,43]

This comet was dubbed NEOWISE because it was discovered by the Near-Earth Object Wide-Field Infrared Survey Explorer space telescope in March 2020. Although the comet could be visible to the naked eye, a viewer might have needed a pair of binoculars first to find it and then to observe the comet's two tails. However, one creation astronomer was able to see the comet even with a bright moon and light-polluted skies.[44]

Comet NEOWISE put on a good show in both the low northeastern pre-dawn sky and the low northwestern evening sky. However, viewing was best in the evening.[45] Comet NEOWISE last approached Earth about 4,500 years ago, which means it could have been seen by biblical patriarchs like Noah![44]

Comets with orbital periods of greater than 200 years are called *long-period comets*, and secular astronomers claim they originate from an enormous reservoir of billions of potential comet nuclei called the *Oort cloud*. However, even the famous secular astronomer Carl Sagan acknowledged that there was zero observational evidence for the Oort cloud's existence.[46]

Secular scientists hypothesize the Oort cloud's existence because they realize that comets typically only remain visible for thousands or tens of thousands of years. After this amount of time, their volatile materials are exhausted, and they can no longer form the long, beautiful tails for which they are famous. Since secular scientists think the solar system is billions of years old, they believe that this reservoir of potential comet nuclei must continually replenish the solar system with new long-period comets. Creationists, on the other hand, point out that an Oort cloud really isn't necessary if the solar system is just 6,000 years old.

The lack of evidence for the Oort cloud is only its most obvious problem. In 2010, secular scientists acknowledged that even if the Oort cloud exists, it can't provide enough long-period comets to keep our solar system "stocked" for billions of years.[47] Perhaps for this reason, scientists have made major revisions to the theory. However, like the Oort cloud itself, this newer version of the theory is highly speculative, with no hard evidence to back it up.[47,48]

Creation by a miracle on Day 4 of the creation week explains how comets even exist, and it does so without the need for wild speculations. Creation only thousands of years ago explains why we can still study comets at all.

References
1. See "An Open Letter to the Scientific Community" on cosmologystatement.org, in which hundreds of scientists express their agreement with the statement "The big bang today relies on a growing number of hypothetical entities, things that we have never observed—inflation, dark matter and dark energy are the most prominent examples."
2. Guth, A. H. 1981. Inflationary universe: A possible solution to the horizon and flatness problems. *Physical Review D*. 23 (2): 347-356.
3. Erickcek, A. L., M. Kamionkowski, and S. M. Carroll. 2008. A hemispherical power asymmetry from inflation. *Physical Review D*. 78 (12): 123520.
4. Inflation was also used to help explain the absence of magnetic monopoles that should have formed when magnetic and electric forces united but which have not been found in the universe. Inflation would hypothetically have stretched these forces out so thinly that they are not observable today. If this were the case, however, then there's no reason why standard elementary particles such as protons wouldn't also be stretched too thin to detect.
5. Thomas, B. We Now Know. *Creation Science Update*. Posted on ICR.org on June 27, 2008, accessed January 16, 2009.
6. Hartnett, J. 2005. *Dismantling the Big Bang*. Green Forest, AR: Master Books, 125. See also Coppedge, D. F. 2007. Inflating the Evidence. *Acts & Facts*. 36 (12): 15.
7. Moskowitz, C. Lopsided Universe Hints at Time Before Big Bang. *Fox News*. Posted on foxnews.com January 15, 2009, accessed January 16, 2009.
8. This series of problems include: a) the singularity; b) energy is converted to matter but not antimatter; c) the ignition of the Big Bang; d) the ignition of inflation; e) the deceleration of inflation; f) compaction of expanding gas into stars; g) ordering of stars into galaxies,

many with rotation; h) ordering of galaxies into superclusters; and i) ordering of matter into planets with unique orbits, rotation axes, and rotation directions.

9. Psalm 19:1-2.
10. Clavin, W. Rotating galaxies galore: New results from ALPINE reveal what appear to be spiral galaxies in the infant universe. *Phys.org*. Posted on phys.org April 21, 2020, accessed April 22, 2020.
11. A familiar example of a Doppler shift is the increase in pitch of an approaching ambulance siren and the decrease in the siren's pitch as it moves away. Light can also be Doppler shifted: light spectra from sources that are moving away are shifted toward the "red" end of the spectrum, and spectra from approaching sources are shifted to the blue end of the spectrum.
12. Thomas, B. Distant Galaxies Look Too Mature for Big Bang. *Creation Science Update*. Posted on ICR.org November 30, 2011, accessed April 22, 2020.
13. Thomas, B. 'Old' Galaxy Found in 'Young' Part of Universe. *Creation Science Update*. Posted on ICR.org May 24, 2011, accessed April 22, 2020.
14. Thomas, B. Distant Galactic Cluster Should Not Exist. *Creation Science Update*. Posted on ICR.org May 21, 2010, accessed April 22, 2020.
15. Thomas, B. Secrets from the Most Distant Galaxy. *Creation Science Update*. Posted on ICR.org November 18, 2010, accessed April 22, 2020.
16. Hebert, J. 2019. Deep-Space Objects Are Young. *Acts & Facts*. 48 (9): 10-13.
17. *The Universe: A Journey Through God's Grand Design*. 2018. DVD. Dallas, TX: Institute for Creation Research.
18. Psarris, S. *What You Aren't Being Told About Astronomy*, vol. 1-3. DVD. Creation Astronomy Media.
19. *Guide to the Universe*. 2016. Dallas, TX: Institute for Creation Research.
20. Williams, J. 2010. *The Work of His Hands*. St. Louis, MO: Concordia Publishing House.
21. Staff Writers. 2017. *Space: God's Majestic Handiwork*. Dallas, TX: Institute for Creation Research.
22. Turner, J. 2019. *Space*. Dallas, TX: Institute for Creation Research.
23. Williams, J. 2018. *The Work of His Hands for Kids*. St. Louis, MO: Concordia Publishing House.
24. Humphreys, D. R. 2005. Evidence for a Young World. *Acts & Facts*. 34 (6).
25. NAM 21: New theory of evolution for spiral galaxy arms. Royal Astronomical Society press release, April 20, 2011.
26. Grand, R. Analysing stellar motions and spiral arm formation in spiral galaxies. Presented at the Royal Astronomical Society National Astronomy Meeting in Llandudno, North Wales, April 17-21, 2011.
27. Russell Humphreys, personal communication, April 25, 2011.
28. Omega Centuri: Colorful Stars Galore Inside Globular Star Cluster Omega Centauri. NASA press release, September 9, 2009.
29. Geller, A. M. and R. D. Mathieu. 2011. A mass transfer origin for blue stragglers in NGC 188 as revealed by half-solar-mass companions. *Nature*. 478 (7369): 356-359
30. Thomas, B. Young Blue Stars Found in Milky Way. *Creation Science Update*. Posted on ICR.org June 9, 2011, accessed October 24, 2011.
31. Tout, C. 2011. Astrophysics: Stars acquire youth through duplicity. *Nature*. 478 (7369): 331-332.
32. Faulkner, D. 2010. Blue Stars—Unexpected Brilliance. *Answers*. 6 (1): 50-53.
33. Gundy, C. NASA's Hubble Finds Rare 'Blue Straggler' Stars in Milky Way's Hub. NASA news release, May 25, 2011.
34. A light-year is a measure of distance, not time. One light-year is about 5.88 trillion miles, the distance that light travels through space in an earth year.

35. Seek Him who "made the Pleiades and Orion" (Amos 5:8)
36. Wolchover, N. Quirky Comet Hartley 2 Confounds Theories on Early Solar System. *Space.com*. Posted on space.com June 16, 2011, accessed June 16, 2011.
37. Thomas, B. NASA Photographs Young Comet. *Creation Science Update*. Posted on ICR.org November 12, 2010, accessed June 16, 2011.
38. A'Hearn, M. F. et al. 2011. EPOXI at Comet Hartley 2. *Science*. 332 (6036): 1396-1400.
39. Thomas, B. Planetary Quandaries Solved: Saturn Is Young. *Creation Science Update*. Posted on ICR.org May 7, 2009, accessed June 17, 2011.
40. Lisse, C. M. et al. 2009. Spitzer Space Telescope Observations of the Nucleus of Comet 103P/Hartley 2. *Publications of the Astronomical Society of the Pacific*. 121 (883): 968-975.
41. See also Humphries, D. R. 2005. Evidence for a Young World. *Acts & Facts*. 34 (6).
42. Hershberger, S. Comet NEOWISE Could Be Spectacular: Here's How to See It. *Scientific American*. Posted on scientificamerican.com July 9, 2020, accessed July 9, 2020.
43. Irizarry, E. For those at northerly latitudes, Comet NEOWISE up in the evening now, too. EarthSky. Posted on earthsky.org July 12, 2020, accessed July 13, 2020.
44. Faulkner, D. A Bright Comet Graces the Morning Sky. Answers in Genesis. Posted on answersingenesis.org July 8, 2020, accessed July 9, 2020.
45. Comet NEOWISE in the Evening Sky. Spaceweather.com. Posted on spaceweather.com July 13, 2020, accessed July 13, 2020.
46. Druyan, A. and C. Sagan. 1985. *Comet*. New York: Random House, 398.
47. Coulter, D. The Sun Steals Comets from Other Stars. NASA. Posted on science.nasa.gov November 23, 2010, accessed July 8, 2020.
48. Thomas, B. New Comet Origins Idea Adds New Problem. *Creation Science Update*. Posted on ICR.org December 9, 2010, accessed July 8, 2020.

Conclusion

Is this a young world after all? Science says yes. The few scientific evidences summarized here represent a vast treasure trove of similar observations. ICR scientists carefully consider and describe them, and present their conclusions and ongoing research on the pages of our free monthly magazine *Acts & Facts*, as well as on our website ICR.org.

Of course, many will refuse to let this evidence challenge their faith in the billions of years claimed by secular scientists. But that is no more the fault of the science than the Pharisees' refusal to believe in Jesus was the fault of the Lord.

In the introduction we asked what difference it would make to discover that real, measurable processes show a young world. It could make an eternal difference for some. It did for a man who told me his conversion story while visiting in my office at ICR one day.

He said that he read Charles Darwin's *On the Origin of Species* just as he entered his teenage years heading into the 1960s. He remembered walking into his living room soon after to tell his parents, "I don't believe in God." Once natural processes replaced the Creator, God became irrelevant.

He then ridiculed all the Christians in his high school. When he entered college, he took a course on the evolution he so proudly hailed. However, his professor admitted some of its problems. For example, instead of the multiple, lower-to-upper fossil sequences that should show step-by-step evolution of one creature into another life form, each of the few and isolated evolutionary transitional form candidate fossils garners differing opinions about its relevance to evolution.

On the last day of class, the professor stated that even though the case for evolution is not as strong as he would like, we have to believe in it because the alternative of creation is "not scientific." At this point, my friend recognized bad logic. Scientists are supposed to uncover all leads, not just close the door to those they dislike. So, finding the evidence for evolution empty, he started searching for different answers.

He came upon the book *The Genesis Flood* by Dr. John Whitcomb and ICR founder Dr. Henry Morris. He said once he read it, everything clicked. If fast water flows made Earth's flat rock layers in rapid succession, evolution had no time. He became convinced of the Genesis Flood and then began reading the Bible.

He understood that if the Bible's history fits the real world, then its spiritual lessons should carry weight. He learned about sin, about sin's awful consequences, and about his need for a Savior to rescue him from his well-deserved death penalty. He repented of his sin and trusted Christ for new life. He will enjoy that relationship forever, as can anyone who repents and trusts Jesus. God used science to gently persuade a young man who once reviled Him. How marvelous!

And for those who already take Genesis at face value, the scientific observations in this booklet confirm that God got His history right. He got everything right. Now more than ever we can affirm the psalmist's words "I will worship toward Your holy temple, and praise Your name for Your lovingkindness and Your truth; for You have magnified Your word above all Your name" (Psalm 138:2). He is worthy of our full trust.

This booklet was adapted from the following materials.

Thomas, B. Earth Hit the 7-Billion Mark Too Late. *Creation Science Update*. Posted on ICR.org October 27, 2011.

Thomas, B. 2016. DNA Trends Confirm Noah's Family. *Acts & Facts*. 45 (7): 15.

Thomas, B. New DNA Study Confirms Noah. *Creation Science Update*. Posted on ICR.org May 16, 2016.

Thomas, B. 2012. Human Mutation Clock Confirms Creation. *Acts & Facts*. 41 (11): 17.

Thomas, B. "80 Million-Year-Old" Mosasaur Fossil Has Soft Retina and Blood Residue. *Creation Science Update*. Posted on ICR.org August 20, 2010.

Thomas, B. 2016. Duck-bill Dinosaur Blood Vessels. *Acts & Facts*. 45 (2): 17.

Thomas, B. Dinosaur Bone Tissue Study Refutes Critics. *Creation Science Update*. Posted on ICR.org November 5, 2012.

Thomas, B. Carbon-14 Found in Dinosaur Fossils. *Creation Science Update*. Posted on ICR.org July 6, 2015.

Thomas, B. Continents Should Have Eroded Long Ago. *Creation Science Update*. Posted on ICR.org August 22, 2011.

Morris, J. 2012. Flat Gaps Between Strata. *Acts & Facts*. 41 (5): 15.

Clarey, T. Cold Slabs Indicate Recent Global Flood. *Creation Science Update*. Posted on ICR.org November 8, 2018.

Thomas, B. Inflation Hypothesis Doesn't Measure Up to New Data. *Creation Science Update*. Posted on ICR.org January 29, 2009.

Hebert, J. "Early" Spiral Galaxy Surprise. *Creation Science Update*. Posted on ICR.org May 1, 2020.

Thomas, B. New Galaxy Model Leaves Old Questions Unanswered. *Creation Science Update*. Posted on ICR.org May 5, 2011.

Thomas, B. New Study Can't Explain Blue Stragglers' Youth. *Creation Science Update*. Posted on ICR.org November 2, 2011.

Thomas, B. Young Blue Stars Found in Milky Way. *Creation Science Update*. Posted on ICR.org June 9, 2011.

Thomas, B. Young Comet Challenges Solar System Formation Story. *Creation Science Update*. Posted on ICR.org June 28, 2011.

Hebert, J. Comet Now Visible to the Naked Eye. *Creation Science Update*. Posted on ICR.org July 15, 2020.